STOP

GLOBAL

WARMING

---

136 WAYS TO REDUCE
YOUR CARBON FOOTPRINT

by Mischa Wu

CLADD
PUBLISHING

Cladd Publishing Inc.
USA

This publication is designed to provide accurate information
regarding the subject matter covered. It is sold with the
understanding that neither the author nor the publisher is
providing medical, legal or other professional advice or
services. Always seek advice from a competent professional
before using any of the information in this book. The author
and the publisher specifically disclaim any liability that is
incurred from the use or application of the contents of this
book.

Stop Global Warming: 136 Ways To Reduce Your Carbon
Footprint

ISBN 978-1-946881-19-9 (e-book)
ISBN 978-1-946881-20-5 (paperback)

CONTENTS

CHANGE STARTS WITH YOU

With the help of everyday people like you; along with government and health organization enforcement, we can turn the burn, into a mass clean air recycling system. Learn to optimize your Green-Efforts and reduce your own Carbon Footprint today with these 136 ways.

By being more mindful, we all can play our part in combating global warming. These easy tips will help preserve the planet for future generations. It all starts with you, right now!

LET'S REDUCE, REUSE, RECYCLE:

REDUCE your need to buy new products or simply use less. This will result in a smaller amount of waste. Consider buying eco-friendly products.

REUSE bottles, plastic containers, glass jars and other items. Reusing water bottles, yogurt cups, bread ties, and other items is being eco-conscious. It will lessen what goes into the landfills, and the amount factories will produce.

RECYCLING unwanted paper, bottles, bags, and more. This is an easy way to make a huge difference. If possible, upcycle tables, furniture, and other outdated items to keep landfills clean. You can recycle almost anything like: paper, aluminum foil, cans, and newspapers.

LARGEST SOURCES OF HUMAN WASTE

Every day, people from around the world are throwing away items that will end up in our landfills. Many in which could have been recycled for reuse. Below is the top 6 places our world's toxic waste is mostly generated from.

#1 Everyday Consumer Waste

- trash or garbage
- paper products
- food debris
- plastic bags
- packaging

- soda cans
- plastic water bottles
- plastic containers
- broken furniture
- grass clippings
- product packaging
- broken home appliances
- clothing
- flooring

#2 *Medical and Clinical Waste*

This is referring to waste produced from health care facilities, hospitals, clinics, surgical centers, veterinary hospitals and labs. They tend to be classified as hazard waste.

- surgical items
- pharmaceuticals
- blood
- body parts
- wound dressing materials
- needles
- syringes

#3 Agricultural Waste

This type of waste includes horticulture, farming, seed growing, livestock, and seedling nurseries.

- empty pesticide containers
- Pesticides
- old silage wrap
- livestock medicines
- used tires
- surplus milk
- cocoa pods
- corn husks

#4 Vehicle Scraps

Old or totaled vehicles are sent to a plant, where all parts are taken out for recycling. The rest is flattened or broken into pieces for recycling. The non-recyclable parts are dumped in our landfills.

- glass
- leather
- textile
- electronics
- plastic
- metal

#5 Construction/Demolition Waste

This is referring to the demolition of old structures to make space for new ones. This is particularly common in cities that are modernizing.

- concrete debris
- wood
- earth
- huge package boxes
- plastics
- old building materials
- insulation
- electrical
- old wires

#6 Electronic Waste

This is waste derived from any electrical devices. These are referred to as e-waste, e-scrap, or waste electrical and electronic equipment (WEEE)

- Devices are contaminated with lead, mercury, cadmium, and brominated flame retardants.

HOW TO PROPERLY RECYCLE

Recycling is done to reduce the need for factories to create more, because we are not reusing what we already have. Recycling uses less energy, lowers air, water and land pollution.

HOW TO PROPERLY SEPARATE YOUR WASTE AND RECYCLE

Paper
- books
- newspapers

- magazines
- cardboard boxes
- envelopes
- scrap paper

Plastic
- plastic bags
- plastic containers or packaging
- water bottles
- rubber bags
- plastic wrappers

Glass
- broken glass
- glass bottles
- glass bowls and dishes

Aluminum
- soda drink
- cans
- foil
- foil cooking containers
- aluminum packaging

136 Ways to Reduce Your Carbon Footprint

HOME

1. **Wrap the Water Heater**
    Wrap your water heater in insulation to save tons of energy and money.

2. **Home Energy Audit**
    Call a home energy audit company to identify areas that are not energy efficient. They will also help you discover ways that you can save energy, based on your unique household needs.

3. **Go Solar**

   Having solar panels installed is becoming more affordable and readily available. Incentives and discounts given by government and energy companies make solar energy something to consider. Solar panels provide energy when the energy companies cannot. It saves you money and gives you peace-of-mind.

4. **Replace Filters on Air Conditioner and Furnace**

   If you haven't changed your air filter, you are wasting energy and breathing dirty air. Cleaning a dirty air filter can save several pounds of carbon dioxide a year. It can also reduce allergies, and asthma to name of few.

5. **Replace Regular Incandescent Light-Bulbs**

   Replace your regular incandescent light bulb with compact fluorescent light (CFL) bulbs. They consume 70% less energy and have a longer lifespan.

6. **Eco-Furniture**

Buying furniture made from wood harvested from sustainable forests, is an excellent idea. Trees absorb carbon dioxide, produce large amounts of oxygen, and provide a natural habitat for animals.

7. **Paint**

The air inside your home is up to three times more polluted than the air outdoors. This is caused by the pollution of off-gassing from paint and finishes. When painting rooms, use no-VOC paint.

8. **Insulate**

Even in new homes, air leaks may account for a 15- 25% loss of energy. Regardless of how we heat or cool our homes, its finding ways to escape.

9. **Buy Energy-Efficient Appliances**

Energy-efficient products can help you to save energy, money and reduce your carbon footprint.

10. **Clean Refrigerator Coils**

Vacuum, dust and wipe with a damp cloth all coils. You'll be surprised at how much grime accumulates.

11. **Replace Old Appliances**

Swapping an old fridge with a new one, could save you thousands over the course of its life. There are many companies that now offer to pick up and recycle your old appliances.

12. **Fridge and Freezer Location**

Placing your fridge and freezer next to the stove, forces the cooling system into overdrive.

13. **Mowing Goats**

If you have a lot of land, buy a few goats to maintain your property. Then you will not have to use the gas or energy guzzling lawn mower that spits out fumes.

14. **Long Grass**

    Let your lawn grow out 2 to 3 inches, instead of cutting it to a golf-course standard. It will grow deeper roots and be more capable of handling droughts. The longer grass blades, will also crowd out weeds and crabgrass, without needed lawn chemicals. Also, leave clippings on the ground to feed the soil.

15. **Get Out the Old Rake**

    Get out your rake and stop using your leaf blower. This will cut down on emissions, gas and energy waste.

16. **Mulch Your Plants**

    Mulch slows water evaporation, restricts weed growth, and adds healthy nutrients to the soil as it begins to break-down.

17. **Cover Your Pool**

    Cover your pool when it's not in use. Not only will it keep the water cleaner, but it will keep it from evaporating, saving you refills.

18. **Tankless Water-Heater**

It's an instant water heater for your entire home. It saves energy by only heating the water as needed, instead of all the time regardless of use.

19. **Green Power**

Contact your power company about your Green power choices. You may be able to purchase solar, geothermal, biomass and wind turbine.

20. **Add Houseplants**

Add one houseplant to each room. Plants purify air from hazardous toxins, breath the CO_2, and replenish oxygen.

21. **Gravel Driveway**

Hard-surface driveways and hardscapes, prevent rainwater from absorbing into the ground. If you must seal your hardtop driveway, avoid coal-tar based sealants, which are nasty to our environment.

22. **Leaky Toilet**
A leaky toilet can waste an amazing 200 gallons of water per day.

23. **Install Ceiling Fans**
Fans use 10% of the energy that your air-conditioning unit does. Fans can reduce indoor temperature by 10 degrees. Also, turn your fan on reverse to circulate warm air downward to recycle existing heat throughout your home.

24. **Programmable Thermostat.**
These devices monitor your homes temperature, so that no energy is wasted on over heating or cooling.

25. **Turn Off the Lights**
Turn of the lights if you are not using them. This is one of the easiest ways to reduce energy, but often times goes unnoticed.

FOOD & DRINKS

26. Eat Naturally

Eating naturally is the best for your health. However, it also cuts down the energy used by factories who produce processed food, along with the plastics used for packaging.

27. Freeze Milk

Buy the largest jug of milk, and only use half at a time. Pour half into a carafe and freeze the rest.

28. Shade-Grown Coffee

Coffee grown in the shade doesn't require the clear-cutting of our rain forests, or the use of pesticides and chemicals.

29. Drink Coffee from A Mug

Chemicals in a Styrofoam cup can leach into your foods and liquids when heated. Most recyclers will not take Styrofoam, so it will end up in a landfill.

30. **Skip Straws**

Disposable straws are simply a waste. They contribute to premature wrinkles around the mouth, are toxic at higher temperatures, and add to our growing waste problem.

31. **Eat local, In-Season Produce.**

When you buy from your local farmer's market, you are receiving food that has not been sent to a factory for processing or packaging. In addition, the food is more nutrient dense.

Also: Eat Less Beef

Cows emit methane into the air, which contributes to global warning. Lower or eliminate your beef intake to help the planet.

32. **Eat Slugs**

They are nutritious, rubbery, free range, easy to hunt, and have been around for thousands of years. It's time to take notice of food supplies that are easy on our planet.

33. **Eat Weeds**

When we did not have farming and a grocery store packed full of ready to eat foods, we ate straight from the earth. This means that things we call annoying weeds today, once was an easy to digest meal.

34. **Vegans Eating for Free?**

Freeganism is an off-shoot of veganism, meaning they avoid all animal products. But the "free" part refers to the free meals they scavenge out of dumpsters. Americans pitch 245 million tons of waste a year, and most of it is edible food.

COOKING

35. **Stop Using Plastic Wrap**

As convenient as it may be, plastic wrap is terrible for the environment.

36. Leftovers To-Go
Keep reusable containers or washed food containers on hand, so that you can send dinner guests home with leftovers.

37. Reusable Cookware
Invest in a quality roasting pan, instead of using a disposable aluminum one. Aluminum is not good for you, or the environment.

38. Manual Can Opening
Switch to a manual can-opener instead of an electric one.

39. Manual Electric Knife
Use a sharp carving knife instead of an electric knife.

40. Cloth Napkins
Use cloth napkins instead of paper napkins. Paper towels produce a lot of wasted energy, factory pollution, as well as large tree consumption.

41. **Skip the Sink**
Dishwashers use half the energy, as hand-washing dishes.

42. **Glass Containers**
Food and beverages stored in plastic containers can contain BPA, a bad chemical for the health of both humans and wildlife. Heating plastic containers in the microwave and washing in them in a dishwasher, causes the plastic to break down faster, allowing them to leach.

43. **No Plastic Water Bottles**
Single-use water bottles are commonly made from polyethylene terephthalate. This causes chemical leaching into your drinking water, and is horrible for your health and the environment. Instead try using a food-grade stainless steel bottle, or one made of BPA-free plastic.

CLEANING

44. **Odor Control**
    Open your window and let clean air in. The inside of your home is more polluted then outside. It is much better for our environment if you refrain from using toxic air fresheners, candles, and plug-in when opening a window will do the trick.

45. **Homemade Cleaners**
    Clean most of your home with a water and white vinegar solution. Store-bought cleaners

contain ingredients that have been linked to thyroid damage, water and air pollution.

46. **Caution Pine & Citrus Cleaners**

Avoid cleaning with pine and citrus on smoggy days. The ingredients can react with ozone to produce cancer-causing formaldehyde.

47. **Vacuum with HEPA Filters**

HEPA filters capture the toxins trapped in your carpets, upholstery and floor. Other filters pull up toxins and spit them into the air we breathe, while containing very little contaminants.

GARDEN

48. **Compost**

Start a compost pile in your backyard. Simply add your food, biodegradable products, and your fireplace ashes to the compost pile? Before long you will have made a natural fertilizer for your garden, lowered the nation's food waste problem; and prevented usable trash from ending up in the landfills.

49. **Water in The Morning/Nights**
Watering during the morning or night will prevent heavy evaporation from the hot sun. The moisture will also penetrate deeper, keeping your lawn healthy.

50. **Natural Homemade Weed-Killer**
Mix undiluted vinegar with a squirt of dishwashing soap, and spray weeds until they disappear. You'll skip the sketchy environmental pollutants in most weed killers, plus, the solution helps keep future generations of weeds at bay.

51. **Build A Wildlife Reserve**
Create these five elements:
- food sources
- water sources
- places for cover
- places to raise young
- sustainable gardening

By planting nectar-producing flowers, and yummy fruits and vegetables, you can easily attract birds and butterflies into your space. This greatly reduces the effects of urban development on wildlife.

52. Use Drip Hoses

Switch to a drip hose. It can be customized by using a hole-punch to target areas that need water the most.

53. Plant a Tree

Planting trees can help in reducing global warming, and cleans the air in close proximity. They not only give oxygen, but also take in carbon dioxide.

OFFICE

54. Online Bill Pay

Pay your bills online, and ask for paperless billing. This will reduce paper usage and postage costs.

55. E-File Your Taxes

File your taxes electronically to cut back on paper usage. Also, ask your accountant about having your tax forms emailed instead of being paper bound.

56. Power Strips

This will curb your electric use, and prevent damage from surges.

57. Delete Old Emails

Did you know keeping your old emails can harm the planet? The more emails you keep, the more server space you use. Data centers are energy hogs.

58. Recycle Old Computers

There are recycling centers for used electronics and computers. Do not allow toxic devices, and batteries to be dumped into our landfills.

59. Refill-Recycle Old Toner Cartridges

Refill and reuse old toner and ink cartridges if you can. They are extremely toxic when dumped.

60. **Try Going Paperless**

Going paperless is not as hard as you think. If you are unable to do so, then print as little as possible. Try emailing documents to yourself, and then using your smartphone to retrieve the email while on the go. Or lower the size of the font, to decrease the paper needed.

61. **Reduce Business Travel via Technology**

Conduct your business meeting with cameras and computers rather than on location? Consider investing in remote meeting tools. You will be saving yourself time, airline tickets, food, labor cost, rental cars, and hotel rooms.

62. **Start an Office Co-Op**

Find other people like yourself who need an office, and office supplies. Or rent a space that utilizes a common area, sharing copiers, fax machines and phone lines.

63. **Invest in Green**

Start investing in sustainable companies. Try finding companies who are committed to reducing their carbon footprint, and improving our planet.

SCHOOL

64. **Take Lunch in a Tupperware**

Each time you throw away a paper sack or plastic bag, more is being produced in a factory for your next purchase. Instead use food-grade stainless steel or BPA-free refillable bottles.

65. **Create a Recycling Center**

A classroom recycling center is the best way to get kids accustomed to being eco-conscious.

66. **Kleenex-Free Free School**

Stop the schools from requiring a huge supply of Kleenex or non-recycled tissue. Same goes for the toilet paper and paper towels used in school bathrooms. If it's not recycled then it came from the clear-cutting of our forests.

67. **Upcycled Art**
Reuse household items to decorate bulletin boards and classroom displays. Create colorful artwork using plastic bottle caps.

68. **Creative Organizers**
Create organizers from common household goods. Use plastic milk cartons to create a student supply center or use cans to create portable supply caddies.

69. **Use Scratch Paper**
Create a special bin for scratch paper that has only been used on one side. Make the students use the used paper first, then the new paper last.

70. **Acid-Free Glue Sticks**
Acid-free glue sticks create fewer messes than liquid glue, and are better for the environment.

71. **Petroleum Free Crayons**
   Most crayons in our schools contain petroleum. There are many alternatives made from soy, vegetable wax, and beeswax.

72. **Go PVC Free**
   PVC is found in many items, and is a major source of phthalates. PVC has been banned from children's products, but is found in including lunch bags, backpacks, and binders. Purchase cardboard binders and natural fabrics to reduce exposure to phthalates. These items will eventually find their way into our landfills, causing the emission of toxins into our air and water supply.

73. **Save A Few Trees**
   Encourage your school to purchase paper products made from post-consumer waste.

ELECTRONICS

74. Batteries

Buy rechargeable batteries. Learn to dispose of old batteries properly.

75. Change Your Screensaver

Complex screensavers, take your computer longer to go into "sleep mode" or prevent it altogether. Choose a simple screensaver and set your computer to "hibernate" quickly after non-use.

76. Standard TV Settings

Select the "standard" picture setting on your TV. Most are brighter than you need and consume up to 20% more energy.

77. Solar-Powered Battery Charger

Charge your cellphone or iPod with a solar-powered battery charger.

78. Video Game Console "Sleep Mode"

Video game consoles consume nearly full levels of power even when turned off. Go to the settings and program your "sleep mode." This will reduce the power it wastes while not in use.

79. **Turn off Electronic Devices**
Turn off electronic devices when you are not using them. Unnecessary usage is wasteful.

SHOPPING

80. **Reusable Shopping Bags**
Using your own reusable shopping bag, makes a huge impact on the reduction of plastics. Reducing plastics in our everyday life is a very important.

81. **Grocery Shop from Home**
Using a delivery service for your groceries can slash your CO_2 emissions by 50%.

82. **Avoid Products with a Lot of Packaging**
Purchase fresh produce, products with less packaging or items in bulk. This will decrease the amount of plastics from packages that end up in our landfills.

GOVERNMENT REFORM

83. Make Polluters Pay

Carbon taxes make polluting more expensive. By punishing those who pollute and giving incentive to those you are energy-efficient, we can make more headway than ever before.

84. Ban Plastic Bags and Light Bulbs

We have promoted alternatives for years, and yet most people refuse to help in saving our planet. By banning plastic bags and energy sucking lightbulbs, we will force everyone to comply.

85. Ocean Tubes

By putting thousands of giant tubes in the ocean, we could use wave motion and a one-way valve to push deep water to the surface. Doing this would bring up essential nutrients to stimulate blooms of tiny marine plants. Subsequently increasing plankton blooms which draw carbon dioxide from the air.

86. **Bury the Carbon**

Scientist could trap the buildup of carbon dioxide in underground aquifers, coal seams or depleted oil and gas fields. It would remove it out of our atmosphere and give us more time to combat the effects.

87. **Fill the Air with Sulfur**

Certain types of aerosols have a cooling effect on the atmosphere. These particles block solar radiation and scatter it back into space. After a volcanic eruption, which can spew millions of tons of sulfur, we find the same type of cooling as aerosols. Some scientists suggest injecting sulfur into the atmosphere to counteract global warming.

88. **Plant Artificial Trees**

It has been proposed building a forest of 100,000 artificial trees to soak up carbon emissions.

89. **Give the Ocean a Dose of Iron**

Tiny photosynthesizing plankton use carbon dioxide to make food. When they die, they sink down to the ocean floor and take all the carbon with them. Since iron stimulates phytoplankton growth, fertilizing parts of the ocean with iron would increase plankton, reducing carbon dioxide in our atmosphere.

90. **Sun Blocking**
Clouds can be created to block the power of the sun. We can begin forming clouds above the ocean by sending salt into the atmosphere.

91. **Collect Cow Farts**
Cow farts and feces release methane emissions into the atmosphere.

DECLUTTER

92. **Don't Fill the Trash**
Instead of throwing away unwanted items
- donate them
- unload them at resale shops
- have a yard sale
- create unique storage ideas

- repurpose them for arts/crafts and school projects
- restore wooden furniture etc.

REDUCE

93. Cut-Off Switch

Install a cut-off nozzle on your shower head. After wetting down, turn off the valve, stopping wasteful water flow while you soap up.

94. Collect Rainwater

Water your lawn and repurpose extra H2O stored by a rain barrel.

95. Reduce Waste

Landfills are a huge contributor of methane and other greenhouse gases. When the waste is burnt, it releases toxic gases in the atmosphere. Reduce the toxicity of our landfills every chance you get.

96. Use Less Hot Water

Buy energy saving geysers and dishwasher for your home. Avoid washing clothes in hot water. Also, avoid taking frequent showers and use less hot water when possible.

97. Quit Junk Mail

You will be saving trees, water and cutting back on pollution.

98. Cancel Subscriptions Being Delivered

Cancel your newspaper and magazine subscriptions and read online instead. There is no need for a forest to die, so that someone can read a newspaper that is also available on their computer.

99. Reduce Water with Low-Flows

About 75 percent of a household's water use comes from the toilet. Outfitting your sink, shower and toilet with low-flow fixtures will conserves water.

REUSE

100. Dry Cleaner Hangers

Take your wire hangers back to the dry cleaners. It helps the business and keeps our landfills cleaner.

101. Buy Used

Before running out and buying something new, check out secondhand stores and garage sales.

VEHICLE & TRANSPORTATION

102. Use Clean Fuel
Support companies that are producing eco-friendly choices.

103. Drive Less or Carpool
By driving less or carpooling, you are reducing the effects of global warming. The largest source of pollution is caused by oil and gasoline.

104. Eco-Cars
If you can't afford an electric car, buy a vehicle that uses the least amount of gasoline possible.

105. Remove roof racks.
Remove storage container, and racks, which reduces your fuel efficiency by 5%.

106. Check Your Tires
Make sure your tires are inflated properly. If they are a little on the flat side, your vehicle

is consuming more fuel which releases more CO_2 in the atmosphere.

107. Don't Skip A Tune-Up

Regular vehicle maintenance will help your car function properly and emit less carbon dioxide.

108. Stop Idling Your Car

Only idle your car if necessary. An average person idles for 15 minutes or more, and longer during colder months.

109. Fly Less

Air travel leaves behind a huge carbon footprint. Consider greener options such as trains, or try vacationing closer to home.

110. ALSO TRY

- Moving closer to work
- Finding a job closer to home
- Walk, jog, skateboard, scooter or rollerblades and skates

LIFESTYLE

111. Use A Clothesline
Most clothes shouldn't be put in the dryer anyway, so this is a great tip for your fabric and the planet.

112. Stairs Are Always Better
Take the stairs instead of the elevator.

113. Time Management
Schedule your errands back-to-back to consolidate your trip and drive less miles.

114. Organic or Fair-Trade Clothes
It takes a third of a pound of pesticides to produce just one nonorganic cotton T-shirt? Buy organic and reduce your carbon footprint.

115. Beeswax Candles
Scented candles release unhealthy chemicals into the air. Instead look for beeswax candles which burn clean.

116. **Become Part of the Global Warming Community**
Connecting with others who share the same concerns, helps strengthen the need for all of us to lessen our carbon footprint.

117. **Become Aware of Your Contribution**
Make it your mission to learn more information about protecting the environment.

118. **Spread the Awareness:**
Always try your best to inform people about global warming. If we just keep this vital information to ourselves, then all of us will suffer the great consequences.

FAMILY

119. **Reusable Cloth Diapers**
If you have a baby, use cloth diapers instead of disposables.

120. **Eco-Crayons**

For your kids, buy crayons made of beeswax, not paraffin. For your babies, try eco-baby shops.

121. **Birth Control**

All humans produce trash, we are the garbage-creation champions. If the average lifespan for a human born today is around 80 years, that's 128,720 pounds of trash per person. If you believe that you can help by curbing the global population, then by all means use birth control.

PETS

122. **Leash Fabric**

Buy your dog a hemp or canvas leash instead of nylon.

123. **Biodegradable Poo Bags**

Don't pick up poo with plastic, and then just toss into the dumpster. Use biodegradable bags which break down easily in landfills.

124. Celebrate Arbor Day and Earth day

It's very important that you not only acknowledge the importance of these days, but also participate. Plant a tree, pick up trash, or join a forum to make a difference. Show others how amazing it can feel to care about our planet.

125. Cut Holiday Waste

We produce 25% more trash between Thanksgiving and New Year's. Be eco-conscious and seek out ways to cut back during these times.

126. Don't Trash the Tree

Recycle rather than trash your Christmas tree. You could also turn it into chips to use in your own yard.

127. Creative Wrapping Paper

Wrap gifts in last year's holiday paper, leftover fabric, newspaper, or paper bags from the grocery store.

128. Lights Off

Turn off or unplug holiday lights during the day, to save energy and make them last longer.

129. E-Cards.

Consider sending a personalized e-card. You can use updated photos of you and your family, pets included. If everyone jumped on this trend, we would save a lot of trees and future trash.

130. Green Gifting

Giving experiences, not things, is a great way to show that you care about them and the planet.

Special Occasions

131. Planning a party or wedding

If you'd like to make a statement with your nuptials, plan a green and organic wedding. Center everything around eco-friendly ideas.

132. Forget the Casket

A modern funeral is normally not environmentally friendly. With large treated boxes lined in toxic finishes that will eventually leak into the soil, it would be wise to at least consider cremation. There are also many other options to review, if cremation is off the table.

PERSONAL HYGIENE

133. Ladies, Use a Menstruation Cup

innovative menstrual cups, are the environmentally responsible choice. Disposable tampons and pads create billions of units of disposable waste each year.

134. Air-Dry Hair

Reducing drying time by even 5 minutes a day saves almost 45 pounds of carbon dioxide emissions annually. Blow drying daily also damages hair and creates dull locks.

135. Re-usable Toilet Wipes

Try using fabric squares that can be washed and reused. After each use put them into a sealed bin and directly empty them into washing machine.

136. **Brushing Teeth the Conservative Way**
If we added up all the water wasted by brushing our teeth, we could give clean water to more than 23 nations. Try wetting your tooth brush and then shutting it off until needed.

WHY GLOBAL WARMING MATTERS

Melting ice-caps, the destruction of vegetation and wildlife, along with violent surges of hurricanes, are all concerning reasons to understand how climate change can affect mankind. Global warming's massive impact on social, economic, and physical health is alarming!

Today we are consumers, and producers, of outright toxic materials and waste. We cannot build landfills fast enough to handle the amount of trash we produce. If it is unable to broken down in our landfills, its either burned or recycled.

It's our duty as citizens of the world, to be adopting a more responsible lifestyle. We do not need to wait for governments or scientist to find a solution. Each individual can be making a difference right now. It's the only practical way to save our planet, before it is too late.

5 KEY POINTS OF GLOBAL WARMING

1. **Rise in Sea Level:** Sea levels are rising in many areas of the world.

2. **Rise in Earth's Average Temperature:** Global temperature has steadily increased during the past 150 years.

3. **Rise in Ocean Temperatures:** The acceleration of vehicles and factories have

resulted in an increase in greenhouse gases getting trapped in the atmosphere.

4. **Shrinking Glaciers:** The glaciers on several mountain ranges, particularly in Greenland and Antarctica, are retreating due to climate change.

5. **Ocean Acidification:** Acid levels in the oceans are increasing. This is due to emission of more harmful gases being absorbed. Oceans are one of our planets natural ways of removing CO_2 from the atmosphere. However, we are producing more than it can handle, causing an over acidification of our oceans.

CO2 IN THE ATMOSPHERE

WHY IT MATTERS: Carbon dioxide roughly makes up 80% of the United States greenhouse gas emissions. Fossil fuels and certain chemical reactions produce an odorless, colorless gas. This gas is responsible for trapping heat in the atmosphere. Despite our planets natural system of removing CO2, levels are the highest they have ever been.

DROUGHTS

WHY IT MATTERS: Droughts occur when there is a major imbalance between evaporation and precipitation. This in turn, causes a prolonged period of dry weather. Droughts have a devastating impact on our health, water and food supplies, animals, and the soil.

Rising Sea Levels (Global Mean Sea Level - GMSL)

WHY IT MATTERS: Our rising sea level is caused by melting ice and a warming ocean. Sea levels continue to rise, leaving coastal areas more vulnerable to flooding and storm surges. This also creates another issue; salt water is now steadily seeping into freshwater aquifers.

Temperature

WHY IT MATTERS: Temperature rise contributes to global climate change. It causes spikes in droughts, typhoons, hurricanes, wildfires, and habitat changes.

SEA SURFACE TEMPERATURE (SST)

WHY IT MATTERS: Our oceans absorb excess heat, causing them to get warmer. This issue affects marine life, decreases healthy fish populations, boost algal blooms, and kills coral. Higher sea surface temperatures increase atmospheric water vapor. SST drives a higher risk of extreme weather events, droughts, and strange storm patterns.

ARCTIC AND ANTARCTIC SEA ICE EXTENT

WHY IT MATTERS: The polar ice caps have existed for millions of years and are reliable proof of climate change. Polar Caps reflect sunlight providing the world with high albedo (reflectivity). This in turn, helps deflect solar radiation, dramatically cooling down our Earth.

Nature Can't Keep Up

Natural carbon sequestration

Natural carbon sequestration is the process where nature removes excess carbon dioxide out of our atmosphere. There are many sources of carbon like: Humans, animals, plants during the night, forest fires, volcanic eruptions, and magma reservoirs. With all of this carbon dioxide being put into the atmosphere, levels must be kept in balance otherwise the surface of the planet would quickly overheat.

Nature designed trees, oceans, soil and some animals to be effective carbon sponges. All organic life on this planet is carbon based. Thus, when plants and animals die, much of the carbon goes back into the ground where it has little impact on global warming.

Nature has a tremendously effective way of dealing with carbon dioxide. However, as the industrial revolution approached, and mankind frivolously used and disposed resources without the environment in mind, our planet become unable to handle the overload.

For example, our oceans have absorbed so much carbon dioxide, they are becoming saturated and acidic. Many tree planting programs have been initiated, but it's going to take a long time before they are mature enough to provide sequestration benefits.

ACTING TO REDUCE OUR CARBON FOOT PRINT ON AN INDIVIDUAL DAILY BASIS, IS JUST AS IMPORTANT AS ACTING ON A LONG-TERM GLOBAL SCALE.

Made in the USA
Columbia, SC
21 December 2018